Written by Claire Jobin
Illustrated by Monique Félix

Specialist adviser:
Marie-Laure Nouel,
International Board of Wool
(Woolmark)

ISBN 0-944589-18-9
First U.S. Publication 1988 by
Young Discovery Library
217 Main St. • Ossining, NY 10562

©*1985 by Editions Gallimard*
Translated by Sarah Matthews
English text © *1987 by Moonlight Publishing Ltd.*
Thanks to Aileen Buhl

YOUNG DISCOVERY LIBRARY

All About Wool

People have very bare skins…

YOUNG DISCOVERY LIBRARY

Fur

Animals have fur, or feathers, or scales to protect them from the heat of the sun, from the cold, and the wind, and the rain.

People have hair, but otherwise their skins are very bare and unprotected. But they do have clever minds, nimble fingers, and the ability to make things.

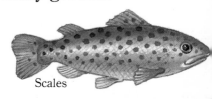

Feathers

If you had been born a hundred thousand years ago, when people were just beginning, what do you think you would have done to cover yourself up against the sun and the wind and the rain, and to protect yourself from prickly plants, and the spiky, scratchy ground?

Scales

Why not use animal skins?

You can eat the meat, and then use the skins to cover yourself. But first the skins have to be cleaned — otherwise they would be very sticky and smelly. Tools were first invented to kill animals, and then to cut off their skins, scrape them clean and sew them into clothes. But this didn't happen all at once — it took a long time, and a lot of hard thinking.

The first tools for working skins were made of flint or bone, which had been trimmed and sharpened.

So as to have enough meat, skins, and bones for tools, people began to keep herds of animals near their homes.

The animals were still wild, but they were becoming used to being near people.

What were these soft white tufts caught on the brambles?
The birds used them to line their nests. They were the hair from wild sheep, which shed some of their fleece in the springtime so as not to be too hot during the summer — this is called **molting.**

The soft white tufts were... ...wool

People began to collect the wool. They rolled it between their fingers and pulled it out into long strands: they had invented spinning. What happened if you joined the strands together, fitting one over the other? This was the first woolen cloth, softer and more supple than animal skins.

Slowly, over thousands of years, wild sheep became the farm sheep of today.

Ewe, ram and lamb

As you can see from this vase, made over 2000 years ago, the Ancient Greeks were very good at weaving.

Two important inventions: the spindle and the loom

Soon, instead of spinning the wool between their fingers, people invented the **spindle**. A spindle is shaped like a long spinning top. A strand of wool from the fleece is attached to the spindle, which is then set spinning as it hangs down. Slowly, as more wool is teased out from the fleece, long strands of **yarn** form and twist around the spindle.

Ancient spindle and woven clothing from northern Europe

Long threads are stretched across the **loom** from one end to the other, then the yarn is threaded through them, over and under, over and under, crosswise. As looms got bigger, weavers were able to make longer, wider pieces of cloth.

An Egyptian loom over 4000 years old

Wool the world over

In the Middle Ages, sheep were herded and wool was woven everywhere. Each country, from the Middle East to northern Europe, had its own specialty. The Spanish raised the best sheep.

Carpets woven in Persia were the finest of all! Woolen cloth from northern Europe was wide and strong.

Merchants traveled far and wide, by sea, on horseback and in camel trains, to trade all these goods.

Sheep hate to be alone.

They need to live in a flock. They are gentle animals, and are easily frightened, but they will follow the leader of the flock anywhere, even into danger...

A large flock is not easy to manage.

Shepherding is skilled work. In the

olden days, a shepherd would spend the year living with his flock. Nowadays, the sheep usually live in fields, and are put in pens in the winter.

One man and his dog

A shepherd's greatest helper is his dog. Sheepdogs learn to round up the sheep, and to separate one from the flock. The shepherd teaches his dog a whole variety of whistles and gestures to tell it which way to go and whether to lie down or to run. Sheepdogs not only help herd the sheep, they guard them too. In past times, they even had to fight off wolves.

In some countries, the fields are too dry to feed the sheep in summer, so the shepherd takes the flocks up into the hills where the grass is fresh and sweet.

So that the sheep don't get lost, they wear bells and bright pompoms.

There are many different kinds of sheep.

Farmers are always trying to improve them. Some ewes which are good for milking are kept so that the farmer can make sheep's milk cheese.

Other sheep provide meat. Their flesh must be tender and taste good.

The best sheep for wool have fine, thick fleeces. Best of all are merinos. There are huge flocks of merinos in Australia, New Zealand and South America. In New Zealand, they say, there are so many sheep that they outnumber the people thirteen to one!

LACAUNE
milk
cheese

ROMNEY
rug wool

MANECH
milk
cheese

HEBRIDEAN
wool for
hand weaving

MERINO
prototype of
wool sheep

SOUTHDOWN
meat

Does wool only come from sheep?
There are lots of other animals all over the world who give wool, but they're not so easily farmed.

Camel

Yak

Camels, yaks and llamas are used mostly to carry things, but they do give wool as well. Some goats with fine hair give a very high quality wool. Mohair comes from the angora goat, while the cashmere goat gives cashmere wool.

Angora rabbit's fur is sometimes mixed with wool.

Llamas and alpacas live in the mountains of South America

What makes sheep's wool so special?

Every hair is curly, and covered with tiny scales. This is what makes wool bouncy and soft.

The scales on a hair

A woolen sweater must be washed carefully so that the scales are not damaged — otherwise the sweater will shrink.

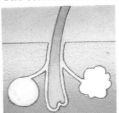

The root of a hair, magnified many times.

On each side of the hair there are little glands producing oil and sweat which make the wool waterproof.

A sheep's fleece has to be well looked after so that it doesn't contain any hard hairs or little insects.

An Australian expert has taken a sheep to one side to examine its fleece.

Sheep are sheared in the spring.
Shearing is done by hand, with scissors, shears, or an electric shearer.

There are many different qualities of wool found in a fleece: it needs grading. This has to be done by hand.

The wool is washed several times to remove any fat, sweat or bits of grass.

To untangle the hairs, the wool is carded with special carding combs. Now it is smooth and ready to be woven.

Can the wool be made even smoother? The finest wool is combed after being carded.

The spindle spun a fine thread. The spinning wheel did it faster. Today's machines spin finer and faster still.

The yarn is put into skeins, and then into balls or onto bobbins. All this is done by machine nowadays.

There is wool of every color of the rainbow. It can be dyed with dyes made from plants or chemicals.

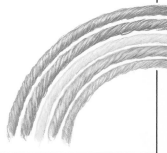

Is wool only used for clothes?

It can be knitted, or woven into carpets, or stuffed into mattresses. It can be made into thread for embroidery or weaving. It can be flattened and beaten to make felt.

But all the sheep in all the world could not provide enough wool to make all our clothes and carpets, stuff our mattresses, make our felt hats and our embroideries!

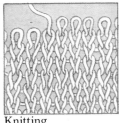

Knitting

Weaving

Felt

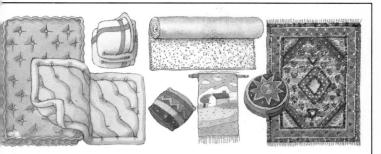

Scientists have invented new yarns made in factories, of polyester and acrylic. But none of them is as good as wool: none is as soft, strong, warm, comfortable and beautiful!

Look at the label on your sweaters or slacks. If it has a mark like this on it,

it means that they are made of pure wool.

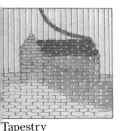

Tapestry

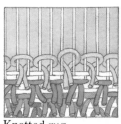

Embroidery

Knotted rug

In the olden days, all knitting was done by hand.

Women would knit night and day. Then, in the sixteenth century, a young Englishman, William Lee, did not want his fiancée to have to work so hard, and invented the first knitting machine.

Nowadays, most knitting is done in factories, on enormous machines.

But you only need two needles and a ball of wool to knit yourself a scarf!

Some knitting patterns are very complicated.

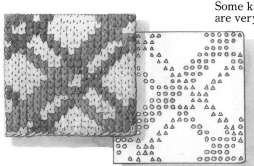

Make yourself a rabbit.

You will need: a pair of old woolen gloves, some bits of wool, cotton wool, a tapestry needle and some scissors.

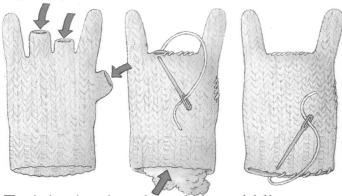

Tuck in the thumb and the middle fingers. Stuff the glove with cotton wool. Sew up the openings.

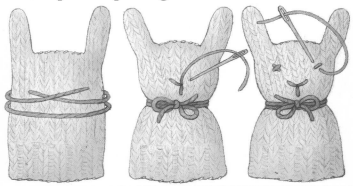

Tie the neck with a length of wool, and sew on eyes and a nose.
What other animals could you make?

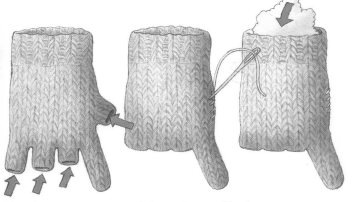

Perhaps you could make a little kitten...

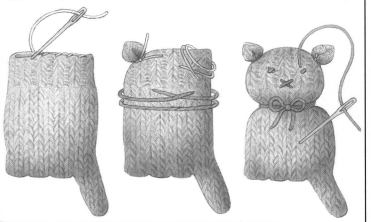

The Happy Sheep

All through the night the happy sheep
Lie in the meadow grass asleep.

Their wool keeps out the frost and rain
Until the sun comes round again.

They have no buttons to undo,
Nor hair to brush, like me and you.

And with the light they lift their heads
To find their breakfast on their beds,

Or rise and walk about and eat
The carpet underneath their feet.

Wilfred Thorley

Index

Alpaca, 23
angora, 22
camels, 22
carding, 29
carpets, 14, 30
cashmere, 22
cheese, 20
cloth, 14
dyeing, 29
ewe, 11
felt, 30
fleece, 10
flocks, 17
knitting, 33
lamb, 11
llamas, 22

loom, 12-13
meat, 20
molting, 10
ram, 11
shearing, 27
sheep, 10
sheepdogs, 19
shepherds, 17-19
skins, 8
spinning, 12
spindle, 12, 29
toys, 34-35
washing, 24, 27
weaving, 12-13
yaks, 22
yarn, 12, 29